Cover illustration: The demands of the Vietnam War were such that Atlantic Fleet (LantFlt) Air Wings joined the rotation during 1965. Among the first was CVW-7 (note the 'AG' tail code, the 'A' indicating Atlantic Fleet). In this view, a 'Jolly Rogers' F-4B (BuNo 151490) is seen on board *Saratoga* in October 1964, prior to its first SEA tour. (USN)

1. An F-4D (s/n 66-7609) of the 366th TFW. The aircraft has taxied straight in from the runway to the disarming pit at Da Nang, September 1970. It is essential that any remaining ordnance be disabled before any other work is done on a returning fighter. The ground crewman under the port wing has a supply of arming pins with their red cloth streamers that he is inserting in the TERs and MERs on the Phantom's pylons. (USAF)

F-4 Phantom

Volume II   ROBERT C. STERN

ARMS AND ARMOUR PRESS
London New York Sydney

Introduction

First published in Great Britain in 1987 by Arms and Armour Press Ltd., Artillery House, Artillery Row, London SW1P 1RT.

Distributed in the USA by Sterling Publishing Co. Inc., 2 Park Avenue, New York, NY 10016.

Distributed in Australia by Capricorn Link (Australia) Pty. Ltd., P.O. Box 665, Lane Cove, New South Wales 2066, Australia.

British Library Cataloguing in Publication Data:
Stern, Robert C.
F-4 Phantom. – (Warbirds illustrated;
no. 46).
1. Phantom (Fighter planes) – History
I. Title   II. Series
623.74′64   UG1242.F5

ISBN 0-85368-828-1

Edited and designed by Roger Chesneau; typeset by Typesetters (Birmingham) Ltd.; printed and bound in Great Britain by The Bath Press, Avon.

The McDonnell Douglas F-4 Phantom II was without a doubt the most important combat aircraft to fight in the skies over Vietnam during the American military involvement in South-East Asia (SEA). US Navy aircraft carriers, including in their Air Wings the first operational Phantoms, began keeping station off the coast of Vietnam soon after the first US military aircraft became officially involved in the Vietnam War, starting with Operation 'Farm Gate' in 1962. It was these Navy Phantoms that were the first of the breed committed to combat, in response to the 'Gulf of Tonkin Incident' in which US Navy destroyers were reportedly attacked by North Vietnamese patrol boats in early August 1964. F-4Bs from *Ticonderoga* and *Constellation* were launched against the port facilities at Vinh on 5 August 1964. Phantoms from the other services that flew the F-4, the US Marine Corps and Air Force, soon followed: Marine F-4Bs arrived at Da Nang in April 1965 in support of 'Leatherneck' land units, and USAF F-4Cs were briefly committed in July 1965, followed in October of the same year by the permanent assignment of Air Force Phantoms to combat in the form of the 8th and 12th Tactical Fighter Wings.

There was often little but frustration in the near-decade that passed before the last Phantom missions over Vietnam were flown in support of departing American ground units. Cold War politics controlled the selection of targets and the frequency of raids, particularly over North Vietnam, to a much greater extent than the demands of military strategy, and months of limited flying would be punctuated by brief periods of frenetic all-out offensive. There was frustration for the Phantom 'jocks', who found themselves frequently at a disadvantage fighting older, slower adversaries in combat rather different from those envisaged by the F-4's designers, at least until improved tactics, training and weaponry gave the Phantom a permanent advantage over NVAF MiGs. Finally, there was the ultimate frustration of leaving a job undone, of feeling less than fully appreciated by the nation whose interests were purportedly being served.

This volume attempts to document in photographs and words the activities and appearance of the F-4 Phantom throughout this period. As with the first volume in this series, it could never have been completed without the help of kind souls at USAF, the McDonnell Aircraft Co. and NASM. In particular, I want to thank Dana Bell and Ron Thurlow, who went out of their way to assist with the collection of photos seen here.

Robert C. Stern

◀2
2. A tractor pushes a 'Chargers' (VF-161) F-4B into a parking spot on *Midway*'s flight deck, Gulf of Tonkin, July 1972. Note the screen cover on the starboard intake; this fitting allowed the running up of an engine on deck without risk of FOD (foreign object damage). (USN)

▲ 3 ▼ 4

3. One of the first Air Wings to be engaged over Vietnam was CVW-7, from the carrier *Independence*. That ship's initial combat tour began in July 1965. This F-4B (BuNo 150482) of VF-41 ('Black Aces') is preparing to fly off upon return to the United States in December of that year. For a change, the underwing racks carry no offensive weaponry. (USN)

4. An F-4E joins up with a tanker on the homeward leg of its mission. Most air strikes required two refuelling rendezvous, inbound and outbound, but, even so, the pair of J79s drank JP-4 so fast that the time over the target was very limited. (USAF)

5. One of the first US Navy Phantoms, an F4H-1 (BuNo 148405, later redesignated an F-4B) of VF-114. The 'Aardvarks' were deployed to the Western Pacific in February 1964 on board *Kitty Hawk*. They completed this tour uneventfully, the last peacetime deployment for a decade. A pair of AIM-7 Sparrow missiles (still missing some of their fins) have been wheeled up prior to being fitted in the specially designed recesses on the underside of the Phantom's fuselage. (USN)

6. On the upwind leg, an F-4B (BuNo 151482) of VF-21 'Freelancers' circles over *Midway*, October 1965, during operations in the South China Sea. The Phantom is armed for an air superiority mission, with AIM-7 Sparrows and AIM-9 Sidewinders. The first engagements with NVAF MiGs came in April 1965, but action was limited through most of that year; by the end of December three MiGs had fallen to Navy fighters. (USN)

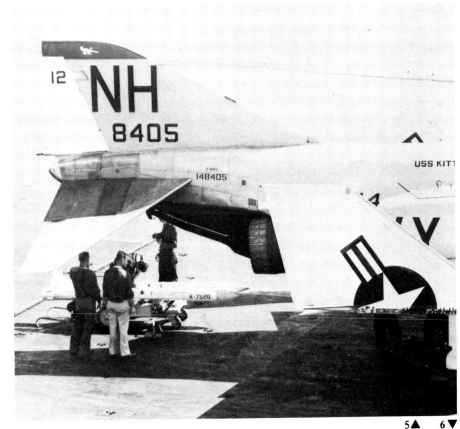

5▲   6▼

7. The first USAF F-4Cs to arrive were part of the 45th TFS, which deployed briefly in July 1965 and then left again. Air Force Phantoms arrived back again for good in October when the 12th TFW moved into Cam Ranh AB. This F-4C (s/n 64-742) is on a photo-chase mission over South Vietnam. It carries a pair of Mk.4 camera pods as well as a quartet of AIM-7 Sparrows in their recessed underbody stations. (USAF)
8. A pair of 12th TFW F-4Cs (64-713 is in the background) formate off a KC-135 over South Vietnam, 6 November 1965. The aircraft are armed for a Skypoint mission with eight M117 750lb GP bombs and the standard four Sparrows in case they have to ditch their bombs and defend themselves. (USAF)
9. An F-4C (s/n 64-730) on the flight-line at Cam Ranh AB in 1965. The camouflage is still the Navy-type (light grey over white, with white control surfaces) carried by all early USAF Phantoms. So far,

the only concession to war in the appearance of this aircraft has been the painting out of the mid-fuselage 'buzz number'. (USAF)
10. This pair of F-4Cs (63-7664 and 63-7623) were assigned to the second USAF unit 'in country', the 8th TFW 'Wolfpack', December 1965. Unlike the Phantom in the previous view, these aircraft retain their 'buzz numbers', the 'FJ' codes carried on the fuselage side. The positioning of large, legible buzz numbers on USAF aircraft came about soon after the Second World War as a result of complaints by civilians about reckless flying by certain young pilots. The idea was that an irate citizen could read the number, a code composed of two letters indicating aircraft type and the last three digits of the serial number, and report the offender. In this case, the buzz numbers were anachronistic and, if not specifically painted out as here, they would disappear when camouflage became standard during 1966. (USAF)

▲11   ▼12

11. The job of recording the results of missions was soon taken over by the RF-4Cs of the 460th TRW, which arrived at Tan Son Nhut near Saigon early in 1966. This photo-Phantom (s/n 64-1054) was obviously flown to Vietnam soon after delivery from the factory. Note the myriad stencilled warnings and notices that almost always disappear shortly after an aircraft reaches squadron service. The deployment of this dedicated reconaissance version of the Phantom freed standard F-4s for normal combat duties. (USAF)

12. The 366th TFW arrived at Da Nang AB in March 1966 and immediately began applying the newly approved SEA camouflage scheme, comprising tan, dark green and medium green, with pale grey undersurfaces. Along with the camouflage came scaled-down national insignia on the fuselage sides and wings and markings of reduced visibility: the large serial number and TAC badge on the tail of the background aircraft have given way to smaller, less conspicuous coding. (USAF)

13. The pilot and his 'back-seater' (often also a rated pilot during the early days of the war) carry out preflight checks on their 366th TFW F-4C (s/n 64-740) at Da Nang AB, 1966. This involved walking around the aircraft, checking all hatches and hoses and making sure all stores were properly mounted. The ultimate decision about the airworthiness of an aircraft always rests with the pilot. (USAF)

▲ 14

14. Almost lost in the patchy low clouds, an F-4C banks towards a target over Vietnam, 13 July 1966. Overcast was a constant problem, caused countless missions to be cancelled and led to the preference on the part of some commanders for radar bombing (Skypoint) missions. (USAF)

15. Afterburners lit, a 366th TFW F-4C lifts off from Da Nang AB, July 1966. (McDonnell Douglas)

16. Inclement weather, particularly during the long monsoon season, left many Phantoms on the flight-line when they would otherwise have been flying. Here, F-4Cs sit in their revetments at Da Nang AB, November 1966, draped in silver tarpaulins and waiting for the weather to clear. (USAF)

17. The monsoon was also a maintenance nightmare, but the demands of the war, particularly for the close support of ground troops, had no respect for the weather. This F-4C of the 480th TFS 366th TFW, at Da Nang AB is being prepared for just such a mission, as can be ascertained from the weapons load, which includes Mk.82 Snakeye bombs (standard 500lb GP bombs with high-drag tailfins for safe low-altitude delivery) and a 20mm gun pod under the centreline. (USAF)

18. The Marine Phantoms of VMFA-115 deployed to Da Nang in April 1966. This 'Silver Eagles' F-4B (BuNo 150429) is seen earlier during weapons-carriage test activities at NAS Patuxent River. (USN)

▼ 15

16▲

17▲   18▼

▲19

▲20 ▼21

19. Another early Marine Phantom unit 'in country' was VMFA-314, which was also based at Da Nang AB in 1966. Marine units en route to combat tours in Vietnam regularly passed through MCAS Iwakuni, Japan, where this 'Black Knights' F-4B (BuNo 149440) was seen. (USN)
20. Once at Da Nang, 'Black Knights' Phantoms began regular combat operations. The nearest F-4B (BuNo 151502) in this view of VMFA-314's flight-line is being prepared for another mission. (USN)
21. An F-4B (BuNo 152236) of VMFA-542 makes a dawn take-off from Da Nang AB in April 1966. Nearly all Marine operations from

Da Nang were ground support missions for the 'Mud Marines' fighting in the nearby jungles. (USN)
22. Two F-4Bs of VF-213 return to *Kitty Hawk* during that carrier's second Vietnam tour, November 1966. The 'NH' tail code indicates CVW-11, and the squadron insignia is a black lion within a white circle. (USN)
23. A 'Black Lions' F-4G (BuNo 150645) returns to *Kitty Hawk*, 1966. The original F-4G (not to be confused with the current Wild Weasel Phantom) was an F-4B modified to carry the experimental AN/ASW-21 data link system. All twelve F-4Gs later reverted to F-4B standard. (USN)

▲ 24

24. An F-4B (BuNo 152312) of VF-32 about to 'trap' on *Franklin D. Roosevelt* in the Gulf of Tonkin, 23 August 1966. 'Swordsmen' markings at the time included a sword down both sides of the aircraft's spine and a white scroll under the Air Wing code, 'AB' for CVW-1, another Atlantic Fleet (LantFlt) wing. This Phantom may have only recently joined the 'Swordsmen', because its markings are incomplete compared to those in the next photograph. (USN)

25. A 'Swordsmen' F-4B (BuNo 152210) photographed on 6 December 1966 gives a 'textbook' look at Navy aircraft markings of the period. The aircraft's number within the Wing, '207', is carried on the nose, the last two digits being repeated on the rudder (and,

on many aircraft, on the inboard flaps). The number '207' means that this is the seventh aircraft of the Wing's second squadron, the yellow fin tip being another indication that the aircraft belongs to the second squadron. The two-letter wing code is carried large on the tail, the last four digits of the Bureau Number (BuNo) are carried on the aft fuselage, and the name of the carrier and the designation of the squadron appear, as is almost always the case, on the aircraft's spine. National insignia are carried on both intakes and on the upper left and lower right wing outer panels. The branch of service is carried on the both sides of the aft fuselage, below the Bureau Number. (USN)

▼ 25

26 ▲

26. Questions about the reliability of the AIM-7 Sparrows carried by Phantoms led to a brief test exercise late in 1966. A detachment of 12th TFW F-4Cs (this one is 64-701, seen on 12 December 1966) were flown to Clark AB in the Philippines, fired off their missiles under test conditions and returned to Cam Ranh Bay within 72 hours. Note the two-letter code carried on the tail, a practice introduced during 1966 to aid in unit identification; 'XD' indicated an aircraft of the 558th TFS. (USAF)

27. During 1967, USAF Phantoms became involved in the strategic war against North Vietnam as well as the tactical support of combat troops in South Vietnam. As a matter of geography, logistics and safety, Air Force units intended primarily to cover targets north of the Demilitarized Zone (DMZ) were based in Thailand. Here RF-4Cs are lined up on the apron at U-Tapao RTAB. KC-135s occupy the apron across the taxiway. (USAF via Dana Bell)

27 ▼

▲ 28

28. Units involved in direct combat support generally maintained an 'alert bird' or two for quick-reaction missions. This F-4C (s/n 64-765) of the 480th TFS, 366th TFW, is armed with a standard mix of short-fuzed Mk.82 'slicks' and AIM-7 Sparrows, June 1967. Note the hose from the APU cart, standing ready to start the two big J79s. This Phantom sports a blue and white rudder flash, seen on many 480th aircraft at this time; it also has the aft-pointing APR-30 RHAW (radar homing and warning) housing above the rudder that

▼ 29

characterized mid- and late-production F-4Bs and Cs. (USAF)

29. An F-4C (s/n 64-762) of the 390th TFS, also of the 366th TFW, drops an empty centreline bomblet dispenser over the A-Shau Valley, South Vietnam, 23 June 1967. The pod might have been dropped accidentally, as such items normally were retained and reused. The rudder flash comprises four stripes of alternating white and blue. (USAF)

30. Another F-4C (s/n 64-803), this time on a mission over North Vietnam, 24 October 1967. Note the fins of an ECM pod just visible beneath the wing tank, indicating the increased threat from SAMs and radar-guided anti-aircraft fire. The tail code and 'Gunfighters' badge on the intake indicate that this Phantom belongs to the 366th TFW. Note the three red stars under the badge, indicating aerial victories, and the 'Roadrunner' cartoon character on the nose. This much decoration on a Phantom was rare in Vietnam. (USAF)

31. Another 'Gunfighters' F-4C (s/n 64-697) over South Vietnam, December 1967. This Phantom carries four Mk.117 750lb bombs and a centreline SUU-16A 20mm gun pod. The SUU-16A did not make up for the lack of any internal gun armament, being only marginally accurate because of vibration, but it was better than no gun at all. (USAF)

▲32 ▼33

34 ▲

32. The winter months brought almost constant solid overcast along with the monsoon rains. Here, an F-4C (s/n 64-701) flies over just such a cloud deck, South Vietnam, December 1967. The 'XD' tail code identified the 558th TFS, 12th TFW, based at Cam Ranh Bay. In the early days of the war, only the first letter of a tail code indicated the squadron, the second letter indicating the individual aircraft. This system was soon superseded by one of consistent two-letter squadron codes, often designed to indicate the unit's base. (USAF)

33. Still carrying six Mk.117s, an F-4C (s/n 64-711), probably belonging to the 366th TFW, banks over South Vietnam late in 1967. The fin tip is International Orange. (USAF)

34. A 'Ghostriders' F-4B (BuNo 151415) recovering aboard *Constellation* in the Gulf of Tonkin after an air superiority mission over North Vietnam, 14 August 1967. It still has a pair of AIM-9 Sidewinders on its underwing rails. The tail markings of VF-142 are light blue and white. (USN)

35. Leaving *Constellation*'s waist catapult, the personal F-4B (BuNo 152250) of VF-142's CO is launched against targets in southern North Vietnam, 26 June 1967. The aircraft is armed with eight Mk.82 Snakeye 500lb high-drag bombs and a pair of AIM-9 Sidewinders. The lightning bolt on the fuselage side is light blue and white, and the 'NK' indicates CVW-14. (USN)

35 ▼

▲ 36   ▼ 37

36. A clear overhead view of a VF-114 F-4B (BuNo 153018), March 1968. The 'Aardvarks' flew off *Kitty Hawk* as part of CVW-11. The fuselage stripe, fin tip and cartoon (from the popular comic strip 'B.C.') are orange trimmed in black. Note the aircraft number and Air Wing code repeated on the starboard wing. (USN)

37. Amid organized chaos, CAG-2 moves into position on the port catapult on board *Ranger*, March 1968. Although it carries the squadron markings of VF-21, the 'Freelancers', this F-4B (BuNo 151478) actually belongs to the CAG, the Air Wing CO, of CVW-2. As far as CAG aircraft go, this one is actually fairly subdued. They frequently carried rainbow-like markings with one colour for each of the squadrons and detachments that made up a Carrier Air Wing. (USN)

38. Another 'Aardvarks' F-4B (BuNo 153045), over the Gulf of Tonkin and returning to *Kitty Hawk* after a MiGCAP mission, 1969. In general, the Phantom's record in air-to-air combat was disappointing, particularly during the preceding year when two Navy F-4s were lost versus one MiG. This led to the setting up of Top Gun, the Navy's fighter school, where intensive ACM (air combat manoeuvring) and fighter tactics training led to a dramatic turnaround in this kill ratio. (USN)

39. A 'Black Knights' F-4B (BuNo 153006) drops its load of eight Mk. 82 Snakeyes from high altitude through solid overcast on a North Vietnamese artillery site just north of the DMZ, February 1968. VF-154 flew off *Ranger* during this tour as part of CVW-2. This is a later B-model Phantom, as can be ascertained from the second, forward-pointing APR-30 RHAW fairing on the tail leading-edge and the additional fairing on the chin infra-red (IR) sensor housing. (USN)

▲40 ▼41

0. Everything hanging in the breeze, an 'Aardvarks' F-4B recovers
n board *Kitty Hawk*, February 1969. The TERs (triple ejector
cks) on the inboard pylons are empty, but the 'standard' self-
efence armament of two AIM-7 Sparrows (in the aft wells) and two
IM-9 Sidewinders is intact, indicating that NVAF MiGs
pparently stayed down this day. In fact, NVAF MiGs pretty much
ayed down for the next three years. After almost daily encounters
ith enemy fighters during 1968, there came a lull in MiG activity
at lasted with little interruption until January 1972. (USN)
. An F-4B (BuNo 152305x) of VF-102 forms up on a tanker.
nlike the Air Force, which uses tailboom-equipped tankers, the
S Navy employs a probe and drogue method of aerial refuelling.
avy F-4s have a retractable fuel probe under the cockpit on the

starboard side, seen extended here. The 'Diamondbacks' first
Vietnam tour in *America* began in May 1968. (USN)
42. The first USAF Phantoms to arrive in Vietnam still wore the
Navy-style colour scheme of light grey over white that characterized
F-4Cs prior to the Vietnam War. This 8th TFW Phantom is seen
over South Vietnam in October 1965, soon after the 'Wolfpack's
arrival 'in country'. (USAF)
43. Soon after the arrival of USAF Phantoms in Vietnam, the grey
paint scheme gave way to improvised tropical camouflage. This F-
4C (s/n 63-7471), probably of the 366th TFW out of Da Nang AB,
is painted in medium green and tan over pale grey. The Air Force
soon formalized the SEA camouflage pattern and colours (which
included a third topside colour – dark green). (McDonnell Douglas)

▲44 ▼45

44. A trio of VF-92 F-4Bs at low altitude prior to recovery aboard *Enterprise*, October 1966. Two of them are dumping fuel from their wing tips in order to reduce landing weight. The tail markings are a yellow chevron and fin tip with a silver chess piece between the letters 'NG', the last indicating CVW-9. VF-92 is known as the 'Silver Kings'. (USN)

45. A look at the hangar deck of *Enterprise*, April 1966. F-4Bs of VF-92's 'Silver Kings' (yellow markings) and VF-96's 'Falcons' (yellow and black fin tips) are being prepared for another strike.

The 'NG' tailcode indicates CVW-9. (USN)

46. An F-4C (s/n 64-735) photo-chase aircraft, at low altitude over South Vietnam, February 1966. The camera pod just visible behind the underwing 370 US gallon drop tank identifies this Phantom's mission. Again, note the painted-out buzz number. (USAF)

47. Contrails form off the wing tips of a 12th TFW F-4C (s/n 64-711) as it bears in on a target during a ground-support mission somewhere over South Vietnam, 1967. The extraordinary humidity frequently caused such contrails at low altitude. (USAF)

▲48

48. The first MiGs fell to Phantoms during 1966. This 8th TFW F-4C (s/n 64-0839) dropping Mk.82 'slicks' through the overcast carries a star on its splitter plate, indicating an aerial victory. (USAF)

49. An early F-4C (s/n 63-7649) of the 18th TFW taxies at Osan AB, Korea, October 1968. It has one more stop before take-off, at the armourer's station (where the arming pins attached to the red cloth strips will be removed). (USAF via Dana Bell)

50. A 'Diamondbacks' F-4B (BuNo 151426) just trapped on *America* – the wire still stretches taut as the Phantom strains forward at full throttle. Steam on the deck indicates that this landing followed shortly after a launch off one of the carrier's waist catapults. (USN)

51. Deck crewmen prepare a VF-33 F-4B (BuNo 151510) for launch

▼49

off the starboard catapult on board *America*. The catapult bridle is attached to two points under the fuselage and then pulled taut. The blast shield has already been raised to protect other aircraft from the J79s' afterburners. (USN)

52. Experience in combat repeatedly drew attention to the fact that the F-4 lacked a gun armament. The answer for the Air Force was the F-4E armed with an internal GE M61A1 20mm cannon, which was rushed through development and testing and delivered to the first combat unit, the 469th TFS, 388th TFW, based at Korat RTAB, in 1968. So rapid was this movement that many units converted directly from the F-4C to the E, skipping the interim F-4D. This view shows an early F-4E (s/n 67-303) over North Vietnam, 30 November 1968. (USAF)

50▲

51▲ 52▼

53. The early models of the Phantom still soldiered on in the original units that first brought them to Vietnam. 'Jeannie' is an F-4C (s/n 64-770) of the 12th TFW from Cam Ranh Bay, seen here on a ground-support mission over South Vietnam, December 1968, and still at war more than three years after first deploying to South-East Asia. (USAF)

54. Well loaded with six Mk.82 Snakeyes on the centreline and six LAU-59A rocket pods under the wings, this F-4D (s/n 66-7475) of the 366th TFW out of Da Nang AB is en route to a Viet Cong target in South Vietnam, February 1969. This is a later F-4D. The housing under the radome, similar in shape to the IR sensor of the F-4C, actually contains an APR-25/26 RHAW set. (USAF)

55. A 'Gunfighters' F-4D (s/n 66-7544) rolling out after landing back at Da Nang AB, its drogue 'chute trailing out behind, February 1969. All the underwing racks are empty, both pylons on each side having been used for ordnance. It carries the large centreline external fuel tank rather than the two 370 US gallon wing tanks. (USAF)

55▼

▲56   ▼57

56. An F-4J of VMFA-232, the 'Red Devils', rolls out after landing on the runway at Chu Lai, 12 April 1969. The F-4J was the Navy's successor to the F-4B, incorporating many of the upgraded systems developed for the D and E models, though not the internal cannon armament. (USN)

57. Loaded down with Mk.82 'slicks', some with fuze extenders to increase the anti-personnel effect, an F-4E (s/n 67-311) of the 388th TFW taxies out at Korat RTAB, late July 1969. The sharkmouth insignia just seemed natural for the redesigned chin of the F-4E and soon became a common marking. (USAF)

58. Drogue 'chute trailing, an F-4D (s/n 66-8702) rolls out at Udorn RTAB, August 1969. The 'OC' tail code identifies the aircraft as being part of the 13th TFS, 432nd TRFW, a unit with a reputation for carrying personal insignia on its aircraft. (USAF)

59. At the instant of launch off *Enterprise*'s waist catapult, a 'Falcons' F-4J dips slightly before rising, early 1969. This VF-96 Phantom is fitted for a MigCAP or BARCAP air superiority mission over North Vietnam, the armament consisting of AIM-7 Sparrows and AIM-9 Sidewinders. (USN)

60. A 'Black Knights' F-4J over the Gulf of Tonkin, 7 April 1969. VF-154 was part of CVW-2 flying off *Ranger* during this period. This Phantom is also fitted for a long-duration air superiority mission, as evidenced by the trio of external fuel tanks. (USN)

61. A 'Silver Kings' F-4J (BuNo 155807) is launched off *Enterprise*'s waist catapult, August 1969. Steam from a launch just moments before off the port catapult swirls around the feet of the deck crewman. (USN)

▲ 62

62. Nose strut still compressed from the rapid deceleration, a VF-96 F-4J 'traps' on *Enterprise*, early 1969. Navy squadron markings evolved continually during this period – note the differences between the tail markings of this 'Falcons' Phantom and those seen in photograph 45. (USN)

63. Another VF-154 F-4J (BuNo 153809), en route to North Vietnam over the Gulf of Tonkin, 10 January 1970. The 'S' on the splitter plate was generally awarded to squadrons with good safety records – a strange prize for a combat unit. (USN)

64. When all else failed, there was always the barrier, a cloth net designed to catch aircraft which for any reason were considered unable to complete a normal arrested landing. Here, a 'Silver Kings' F-4J (BuNo 155807) catches the barrier on board *America*, 21 July 1970. (USN)

65. Crews begin the walk back to the squadron building after a mission, Phu Cat AB, South Vietnam, March 1970. The 'HB' tailcode identifies this F-4D (s/n 65-789) as belonging to the 37th TFW. Ground crew have already started preparing these Phantoms for refuelling in anticipation of the next sortie. (USAF)

▼ 63

64 ▲   65 ▼

▲ 66

66. A pair of 'Wolfpack' F-4Ds form up under a KC-135 outbound from Ubon on a ground support mission over South Vietnam. The doors covering the refuelling port behind the cockpit have been opened preparatory to taking on fuel. (USAF)

67. 'Arizona Chicken', a beautifully decorated F-4E (s/n 66-371) of the 34th TFS, 388th TFW, out of Korat RTAB. The aircraft is off the wing of a KC-135 en route to South Vietnam, 28 May 1970. It is armed with a full load of a dozen Mk.82 'slicks', those on the wing pylons fitted with 36in fuze extenders. The 'chicken' in question is actually a rendition of the 'Roadrunner' cartoon character. (USAF)

68. Artwork flourished in many Phantom units, particularly those based in Thailand, up until 1970. 'Easy Rockin Mama II' was a 34th TFS, 388th TFW F-4E seen here prior to a mission at Korat RTAB, fully loaded with Mk.82 slicks. (Ron Thurlow)

69. A close-up view of the 'artistic' part of 'Easy Rockin Mama II's markings. The background and lettering are two different shades of blue, the chair and Tommy Gun are shades of brown and grey, and the remarkably proportioned young lady has blonde hair, blue eyes and a suntan. (Ron Thurlow)

▼ 67

68▲   69▼

39

70. An F-4E (s/n 67-269) of the 34th TFS, 388th TFW, sits in its revetment at Korat RTAB armed for an interdiction mission, November 1970. Three Mk.82 'slicks' are visible on the inboard pylon and a pair of CBUs (cluster bomb units) on the outboard. (USAF)

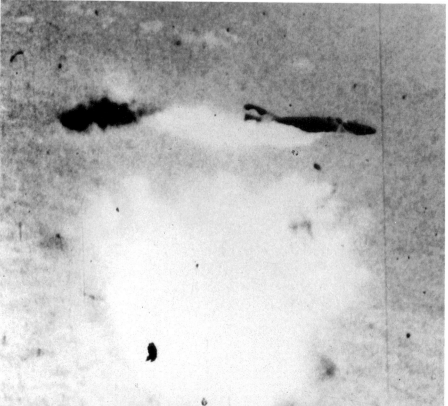

▲71 ▼72

73▶

71. 'Honky Tonk Angel' , an F-4D of the 13th TFS, 432nd TRW, seen at Udorn RTAB. This was one of the few (and possibly the only) sharkmouthed Ds ever photographed. Note the ECM pod, the fuze extenders on the Mk.82s and the formation tape lights on the nose and fuselage side. These were later designed to give off a faint glow to aid night-time formation flying. (Ron Thurlow)

72. The intense SAM and AAA threat over North Vietnam made the carriage of ECM pods a necessity. In general, they succeeded in defending the Phantoms, but the occasional failures were tragic and spectacular – witness the demise of an RF-4C over Hanoi after being hit by a SAM-2, 1971. (USAF)

73. Sunlight reflects off flooded rice paddies behind an F-4D (s/n 65-785) of the 'Triple Nickels', the 555th TFS, 432nd TRW, banking for home after dropping its ordnance on suspected VC (Viet Cong) positions in the patch of forest in the background. (Ron Thurlow)

74. A 'Sundowners' F-4B climbing vertically for the photographer while working up for *Coral Sea*'s 1972 Vietnam cruise. The sharkmouth and the setting sun on the rudder were red and white. VF-111 traditionally had some of the most colourful markings among Navy squadrons. (NASM)

75. An F-4D of the 43rd TFS, 8th TFW, the 'Wolfpack', sits in its revetment at Ubon RTAB, 10 October 1969. A pair of 2,000lb KMU-351 Paveway laser-guided bombs are on display in the foreground, broken down into their major components. The bomb's body was a standard Mk.84. The flexible nose fitting, here still missing its guidance vanes, contained the laser seeker. The tailfins of this early model Paveway were small and fixed. Later models had larger, sometimes folding fins for better aerodynamic performance. (USAF)

76. 'Little Chris' was a brightly marked, sharkmouthed F-4E (s/n 68-322) of the 388th TFW, and is seen here at Korat RTAB in 1970. The Phantom carries a pair of ECM pods under its wings. (USAF)

◀**74**

75▲   76▼

▲77

77. An F-4D (s/n 65-677) of the 433rd TFS, 8th TFW, from Ubon RTAB, off the wing of a KC-135 and target-bound on a bombing mission over North Vietnam, May 1970. The target probably is a bridge or other similar high-value site – as can be surmised from the pair of 2,000lb KMU-351 laser guided bombs carried underwing inboard of the 370 US gallon external fuel tanks. Another Phantom in the same flight would carry a laser designator pod, such as Pave Knife. (USAF)

78. A VF-111 ('Sundowners') F-4B (BuNo 151489) about to launch from *Coral Sea* against targets in North Vietnam with a full load of Mk.82 'slicks' and AIM-9 Sidewinders, April 1972. The big centreline external fuel tank carried 600 US gallons. (USN)

79. Just having landed aboard *Saratoga*, the CAG aircraft of CVW-3, an F-4J (BuNo 157293), strains against the wire. The squadron emblem on the intake in yellow, black and white is 'Felix the Cat', one of the Navy's oldest insignia, first employed by VB-2B in 1928. (USN)

80. The pace of missions slowed considerably after the signing of the Paris Accord on 27 January 1973. With the last 60,000 US troops being withdrawn from South Vietnam, air strikes from bases in Thailand or from aircraft carriers offshore became the sole option available to support the faltering South Vietnamese government. Here a trio of F-4Bs of VF-151, the 'Vigilantes', from *Midway*, are seen over the Gulf of Tonkin, March 1973. (USN)

▼78

▲81　▼82

81. A late F-4E (s/n 69-261), probably of the 13th TFS, begins to taxi under the watchful gaze of its groundcrew, Udorn RTAB, 15 August 1973. The significance of the black widow spider with the number '421' on the splitter plate is unknown. That this is a late model E can be ascertained from the extended gun sleeve under the nose, designed to prevent the ingestion of muzzle gases. (USAF)
82. A crowd of Phantoms (as well as A-6s and A-7s) forward on the flight deck of *Midway*. CVW-5's last war tour ended in March 1973. The F-4Bs are from the 'Vigilantes' of VF-151 (yellow tails with black stripes) and the 'Chargers' of VF-161 (black tails with red lightning bolt). (USN)
83. Now sporting a solid red tail, an F-4J (BuNo 157265) of the

'Red Devils' of VMFA-232 rolls out at Cubi Point, May 1971. Later F-4Js had DECM (deceptive electronic countermeasures) sets in semi-recessed housings mounted high on the jet intakes on both sides, visible here above the national insignia. (USN)
84. Just making sure nothing bad happens to him, a pair of VF-213 F-4Js escort a Tu-20 'Bear' reconnaissance aircraft over *Kitty Hawk*, 7 May 1971. The Russians took a lively interest in the activities of the Seventh Fleet and regularly overflew 'Yankee' and 'Dixie' stations. 'Bear' crewmen and their escorting Phantom 'jocks' frequently exchanged internationally recognized hand signals during these encounters. (USN via Dana Bell)

85. An RF-4C (s/n 68-609) of the 14th TRS, 432nd TRW, out of Udorn RTAB between solid cloud layers over Vietnam, 12 August 1971. This is a later model photo-Phantom, with RHAW antennas on the nose camera fairing and a LORAN (long range navigation) 'towel rack' on the spine. (USAF via Dana Bell)

86. 'Trapper', another 'Triple Nickel' F-4D (s/n 66-7554), seen at Udorn RTAB, 21 July 1971. The fin tip and corresponding stripe on the nosewheel door are dark green, and note the two red victory stars on the splitter plate and the 'Peanuts' cartoon character 'Snoopy', complete with flying helmet and red scarf, on the intake. (USAF)

85▶

▼86

89▲

87. An F-4J (BuNo 155800) in the markings of the
'Falcons' of VF-96. The six stars along the fuselage,
each in a different colour, stand for the Wing's
component units. This aircraft (which also appears in
the background in photograph 88) is doubly
important because it was also the mount in which
Lts. Randy Cunningham and William Driscoll got
their third, fourth and fifth MiG kills on 10 May
1972, becoming the first aces of the Vietnam War.
Outbound from Hanoi, the F-4 was hit by a SAM and
had to be ditched. Cunningham and Driscoll were
rescued. (USN)
88. An F-4J (BuNo 155799), the personal mount of
the CO of the 'Silver Kings', VF-96, is positioned on
one of *Constellation*'s waist catapults, May 1972. The
catapult bridle has yet to be attached but the blast
barrier has already been raised in preparation for
launch. (USN)
89. A 'Tomcatters' F-4J of VF-31, having completed
a sortie, taxies across *Saratoga*'s bow into a parking
position, Gulf of Tonkin, August 1972. The tail
marking and fuselage flash are red, white and black,
'AC' indicating CVW-3. (USN)

90. A self-contained 'laser bomber', an F-4D (s/n 66-7680) of the 433rd TFS, 8th TFS, waits its turn to taxi at Ubon RTAB, 12 May 1972. Visible are a pair of 1,000lb KMU-353 'smart' bombs and a Pave Knife laser designator pod. Only six 'Wolfpack' Phantoms were modified to carry the Pave Knife pod; one pod could, however, designate targets for several aircraft. (USAF)

91. A 'Wolfpack' F-4D (s/n 66-8803), target-bound from Ubon RTAB and loaded with KMU-353s, 6 June 1972. Note the LORAN 'towel rack' antenna, which allowed precise navigation over long distances. (USAF)

92. The same Phantom again, with the designator aircraft, another F-4D equipped with a Pave Knife pod, now coming into the scene from the left. (USAF)

93. Armed with two AIM-7 Sparrows, two KMU-353s and a pair of ECM pods, an F-4E (s/n 69-266) of the 334th TFS, 4th TFW, flies over South Vietnam on its way north, 10 July 1972. This unit was normally based in the USA at Seymour-Johnson AFB, but it forwarded aircraft TDY (temporary duty) to the 8th TFW at Ubon RTAB for the 'Linebacker I' offensive (which explains the TAC badge on the tail). (USAF)

▲94   ▼95

94. A 'Night Owls' F-4D (s/n 66-8722) of the 497th TFS, 8th TFW, taxies out to the arming pit at Ubon RTAB. It is armed with second-generation Paveway 'smart' bombs, distinguished by their steerable tail fins. The owl insignia on the intake is black and white. This unit specialized in night operations. (USAF)

95. A LORAN-equipped F-4D (s/n 66-8738) rolls away from the flight-line at Ubon RTAB, September 1972. Forwarded 4th TFW aircraft can be seen in the background, identified by their 'SJ' tail codes. (USAF)

96. A TDY Phantom forms on the wing of a TDY tanker, September 1972. The Phantom is an F-4D (s/n 65-635) of the 334th TFS, 4th TFW, out of Ubon RTAB; the tanker is a propeller-driven KC-97 of the 134th ARG, an Air National Guard unit activated for 'Linebacker I'. (USAF)

96 ▼

◄98

99▲

97. (Previous spread) A dramatic view of a pair of 'Wolfpack' F-4Ds en route to Vietnam during 'Linebacker I', September 1972. The nose of a third Phantom is coming into the scene from the left. The nearer Phantom is 66-234 of the 435th TFS; further away is 66-8705 of the 433rd. This upward view gives a good idea of the recessed 'wells' for the AIM-7 Sparrows. (USAF)

98. The same 'Wolfpack' trio. The third aircraft is another F-4D (s/n 66-8795) of the 25th TFS, the third component squadron of the

8th TFW. The fin tips are (from front to back) red; yellow and black checkerboard; and dark green. (USAF)

99. An F-4E (s/n 69-293) of the 4th TFW taxies out at Ubon RTAB, September 1972. It carries an impressive weapons load – six Mk.82s and four Mk.84s, totalling 7,000lb of bombs – as well as at least one ECM pod. Note the Wing badge on the intake as well as the TAC badge on the tail. (USAF)

100. 'Wolfpack' Phantoms begin the long journey to work. (USAF)

100▼

▲101 ▼102

103 ▲

101. A sharkmouthed F-4E (s/n 68-336) of the 421st TFS, armed for an air superiority mission, taxies out to the arming pit at Takhli RTAB, September 1972. It is fitted with two ECM pods, two AIM-7 Sparrows and four AIM-9 Sidewinders. The AIM-9s have orange rubber caps over the IR seeker heads to prevent damage by the sun; these will be removed, along with all the arming pins, before take-off. (USAF)

102. A pair of fully loaded F-4Ds from the 388th TFW head north over the Gulf of Tonkin at the height of the 'Linebacker I' offensive, 30 September 1972. The aircraft are each loaded with twelve Mk.83 1,000lb 'slicks' and a pair of ECM pods. 'Linebacker I' coincided with Richard Nixon's bid for re-election and an attempt to force the North Vietnamese back to the negotiating table. The climax was Operation 'Prime Choke', aimed at severing the supply routes between China and Vietnam north of Hanoi, an area previously off-limits because of the proximity of the Chinese border. (USAF)

103. 'Linebacker' brought some of the older F-4Cs back to Vietnam, such as 64-963 of the 49th TFW, normally based at Holloman AFB but now operating out of Takhli RTAB. (USAF)

▲104

104. Takhli RTAB had been closed down prior to 'Linebacker I', which began in May 1972, when it was reopened to accommodate the F-4Cs of the 49th TFW. These aircraft were replaced late in the year by F-111s of the 474th TFW from Nellis AFB. During the transition, the apron got rather crowded. (USAF via Dana Bell)
105. An F-4D (s/n 66-269) of the 13th TFS, 432nd TFW, out of Udorn RTAB and on its way north, October 1972. Besides four AIM-7 Sparrows, this Phantom carries a pair of ECM pods. The nearer pod is the small, first-generation version, the other is the new QRC-335 of greater sophistication and power. (USAF)
106. An F-4D (s/n 66-8737) returns to Udorn RTAB, October 1972. The 'UP' tail code identifies the 388th TRW. The fin tip and cockpit trim are light blue. (USAF)

▼105

107. One of the characteristics of the final phase of US involvement in Vietnam was an increasingly desperate desire to find a 'magic' solution. A manifestation of this was the bombing of VC 'sanctuaries' in Cambodia and, later, Laos. Here, an F-4E (s/n 68-498) of the 34th TFS, 388th TFW, from Korat RTAB, drops a pair of Mk.84 2,000lb bombs from above the clouds on targets in Laos, early December 1972. At least one CBU-58 cluster bomb remains in place. The aircraft partially visible in the foreground belongs to the 25th TFS, 8th TFW, from Ubon. (USAF)
108. Another Skyspot mission over Laos, early December 1972. The F-4E, dropping a full load of Mk.84s, is part of the 388th TFW based at Korat RTAB; the D is part of the 13th TFS, 432nd TRW, based at Udorn. (USAF)

106 ▲

107 ▲   108 ▼

109. Deck crewmen crouch in the safe space between the catapults as a 'Screaming Eagles' (VF-51) F-4B (BuNo 151475) is given the launch signal on *Coral Sea*, September 1973. The tail markings are black with red trim and lettering. The deck crew are more colourful, with yellow, red, green, and black and white checkered shirts indicating their activities and helmet markings indicating authority. On a noisy carrier deck, amid the swirling steam of catapults, high visibility and instant recognition are essential. (USN)

110. A late-war mission from Ubon by an F-4D (s/n 66-249) of the 8th TFW, 4 August 1974. By this time no-one had any illusions about winning the war and most people were just going through the motions. Note the absence of any tail code and the irreverent grafitti on the intake. (USAF)

◀ **109**

110 ▼

111. A new feature added to many Air Force Phantoms during 1973 was TISEO (Target Identification System, Electro-Optical), a fancy name for a TV camera with a telephoto lens. TISEO was mounted on the port wing leading edge, feeding images to the WSO's cockpit displays. Radars can spot and missiles can track targets well beyond unaided visual range, but, even with IFF, caution required that no target be attacked until it could be visually identified as non-friendly. TISEO, by extending the range of visual recognition, extended the distance at which missiles could be fired. Given the closing speeds of modern combat jets, being able to launch even a few seconds earlier could make the difference between engaging with missiles and becoming involved in a dangerous dog-fight. (USAF)

112. An F-4D (s/n 66-679) of the 34th TFS, 388th TFW, taking off from Korat RTAB. Another feature of the late war period was the gradual reorganization of the Air Force back to a peacetime footing. CONUS or USAFE units were rotated back to their bases, leaving only PACAF to carry on. (USAF)

113. Incredibly loaded, an F-4E (s/n 67-309) of the 388th TFW heads out on one of the war's last missions, November 1974. It is carrying ten Mk.82 'slicks', a pair of CBUs, a rocket pod and a pair of AIM-7 Sparrows! (USAF)

114. War's end. An RF-4C (s/n 68-596), probably of the 432nd TRW (note the stencilling on the ground crewman's cap) is directed to a parking spot on the ramp at Hickam AFB, Hawaii, one stop on the way home. (USAF)

▲111 ▼112

113▲ 114▼

▲ 115

115. The 388th TFW became the 3rd TFW and had its home base established at Clark AFB, the Philippines, while it continued to fly occasional exercises out of Korat to keep the attention of the Vietnamese. This pair of 3rd TFW F-4Ds (led by aircraft 65-755) were back at the old base for Exercise 'Commando Scrimmage', 11 March 1975. (USAF)

116. These sorties were indeed an exercise in futility. The NVA offensive that would overwhelm the last RVN resistance was well under way and, as long as this 3rd TFW F-4D (s/n 66-234) carried no ordnance, it stood no chance of influencing the battle – a sad but perhaps fitting end to a conflict that was debatably noble but undeniably brave. (USAF)

▼ 116